T0194974

Praise for Salem Radio

"I agreed to air *All the Difference* on a dare. Convinced few would listen, I laughed as we signed that first short-term contract. The Salem Radio staff also chuckled, knowing what I did not – that radio ministry is powerful. From the start, we were astonished at the impact. *All the Difference* now gets mail every day, engaging with people eager to reason from the Scriptures. 5% of our audience are non-Christians, a majority are under 50 years old, and many share about what God is doing in their lives. We remain humbled and grateful to Salem for making 'all the difference' for us."
—**Wayne Braudrick**, lead pastor, Frisco Bible Church, Frisco, TX

"To all who may be considering starting a program on Salem Radio. Reformed Heritage's program, *Abounding Grace*, has been airing on KFAX for about six years, and we couldn't be more pleased with the results. Approximately 85% of our visitors come from listeners. About 65% of our current members have come as a result of the program. Christ has also used radio to bring sinners to a saving knowledge of Himself. RHC is a church with people who love to evangelize, but the radio program is still our main source of reaching the unchurched and the lost."
—**Pastor Gary Wagner**, Reformed Heritage Church, Los Gatos, CA

"Nearly every Sunday, I meet someone who is visiting our church because they heard one of our messages on the air. For the past five years, our daily radio ministry, *Fully Devoted* on Salem Radio has been an outreach tool that consistently impacts our community and builds our congregation."
—**Jesse Smith**, senior pastor, River City Christian, Sacramento, CA

“We have been a part of the Salem broadcast family from the time they began on WLQV AM 1500 in Detroit. It’s amazing how often we meet people at ministry events and sometimes even in the grocery store, who tell us how much they love listening to our broadcast. Partnering with WLQV has made it possible for us to reach southeastern Michigan with God’s Word. It is always a blessing when a visitor joins us in one of our church services and tells us they heard *Kingdom Living* on WLQV. Thank you for giving us a stronger voice!”

—**Pastor Mark Byers,** Calvary Christian Church and *Kingdom Living*

“Verse by Verse radio has been a part of WTBN radio for many years. Not only have we heard of people coming to faith in Christ as a result of these broadcasts, but hardly a week goes by that I don’t meet someone who is visiting our church because they first heard us on the radio. In addition, it is very common to find that many of our new member classes consist of people who were first introduced to our church by the radio messages. I am so thankful for Salem Radio and the outreach of our pulpit ministry through their ministry.”

—**Steve Kreloff,** pastor of Lakeside Community Chapel in Clearwater, FL

“*The Full Gospel Hour* has been an important ministry for Bethel Gospel Tabernacle for more than 30 years on WMCA and for over 60 years altogether. Our radio program helps to expand the boundaries of our church and its ministries. It reaches those who may not have a church or are looking for one. I encourage churches and ministries to expand the reach of their church for Christ through a Salem Radio ministry.”

—**Bishop Roderick Caesar,** Bethel Gospel Tabernacle, Jamaica, NY

"We are not a mega church, but radio has given us a mega ministry. We've found that broadcasting weekly on 570 (Salem Radio WMCA) allows us to be an Acts 1:8 church. We are witnesses through this powerful medium both in Jerusalem (cross city), Judea (cross community), Samaria (cross culturally) and to the ends of the earth (cross continentally) we regularly have visitors from the radio broadcast."
—**Pastor Dave Watson**, Calvary Chapel, Staten Island, NY

"Before joining Salem Radio and the WMCA broadcast family, Donna Baptiste Ministries (DBM) was a little-known ministry in the Metro New York City area. Now, as a result of WMCA, DBM is better known and has caused doors once closed to open to the ministry of DBM."
—**Pastor Donna Baptiste**, founder, Donna Baptiste Ministries, Brooklyn, NY

"I can't put into words how grateful I am for Salem Media! As a pastor, and especially as a Chaplain for our County Sheriff's Department, when I found out in 2013 about how prevalent the 'sex slave industry' is in the United States, I knew I had to do something! I got an idea about doing a radio program about human trafficking here in America. I pitched the idea to our local Salem station and they immediately went to work to help me get the program on the air!!! We are in our 4th year now and it has been remarkable the growing awareness all across our country. I thank God for Salem Media giving me an opportunity to share with our area, and with the world over the internet, a great burden."
—**Chuck Winters**, pastor of Cedar Heights Baptist Church, North Little Rock and Host of PATH Saves

"I am so thankful for my relationship with Salem Media. I host a weekly radio program in Central Arkansas called *Truth on the Go with Andrea Lennon.* The weekly program has resulted in more social media followers, increased book sales, and new speaking requests. In addition to my weekly program, The Fish runs one minute *Moment of Truths with Andrea Lennon.* A week doesn't go by that I don't hear from listeners who are encouraged by these short messages of God's hope, peace, and power. Radio is a powerful mode of communication and I'm excited to continue to use it to reach more women with the gospel of Jesus Christ."

—**Andrea Lennon**, True Vine Ministry in Conway, AR and host of *Truth on the Go with Andrea Lennon*

"Bethel Baptist Church has been partners with Salem Broadcasting (FaithTalk 99.5) now for three years. When we first entered into this partnership we had no idea how to do a radio program, much less one that was a live, remote broadcast. Needless to say, we were pretty nervous about it, but the people of Salem Broadcasting guided us each step of the way. They not only helped us get started, but they have been there for us each time we needed assistance. Salem Broadcasting is not just a 'business' partner, they are a vital part of the ministry team of Bethel Baptist Church in helping us to proclaim the Gospel message."

—**Dr. Joe Manning**, pastor of Bethel Baptist Church, Jacksonville, AR and host of *Apologetic Cafe*

"With a famine in the land not of bread nor of corn but of the Word of God, WHKW-AM (Salem Radio) has been that of Joseph's granaries. A place to feed the Lord's flock."

—**Pastor Jeff Tauring**, Liberty Valley Church, Macedonia, OH and host of *Time in The Vineyard*

"Through WAVA 105.1 FM Salem Network, the Reaching Your Heart Broadcast reached into the White House and touched a secret service staff member with the love of Jesus. I baptized his family into Christ, and he gave Bible Studies at the White House because of the broadcast's influence. Matt will tell you that Christian radio saves lives for eternity.

—**Pastor Michael Oxentenko**, Reaching Hearts International Church: "Reaching Your Heart" on WAVA FM and SXM Family Talk

"Christian Salem Radio stations like WAVA and SiriusXM Family Talk give us a platform to reach tens of thousands of people with Biblical teaching. It helps us generate hundreds of web site hits and phone calls weekly, and see lives impacted daily for Jesus Christ."

—**Angel Voggenreiter,** director of *So What* radio ministry with Pastor Lon Solomon

"Salem Radio has allowed us the opportunity to extend our ministry well beyond the walls of our local church in a way that enables the Gospel to impact the lives of people across the nation that, perhaps, other forms of new media may be leaving behind. We've heard from listeners in many walks of life, including commuters, the elderly, prisoners, and the homeless about how our radio program has directly impacted their lives, and enabled them to grow deeper in their faith."

—**Pastor Gary Hamrick**, Senior Pastor, Cornerstone Chapel, Leesburg VA

"During our WAVA ministry program on Salem Radio, we promoted a new mid-week Bible study, encouraging listeners to come out to the church. Out of the 30 people that came out, 28 were directly from WAVA."

—**Pastor Pitts Evans**, Whole Word Fellowship, Oakton, VA and host of *Jesus In Your Morning*

"'We've been listening now for a while and we wanted to come meet you. We've been looking for a church and we found you on the radio.' I hear this kind of thing all the time after one of our four Sunday services. We've broadcast *Life Lessons* on the Salem Media station KGNW, 820AM The Word, for over 14 years. You want to widely preach the Word, radio is the way to do it."

— Dr. **Steve Schell**, senior pastor, Northwest Church, Federal Way, WA and host of *Life Lessons*

"In 2010, Living Faith Ministries (LFM) opened its doors in Clinton, Maryland and stepped out in faith to begin our journey. The mandate was simple, spread the gospel and fulfill The Great Commission. God orchestrated a chance meeting with the Salem Radio station WAVA. We didn't know anything about broadcasting nor did we have an audio ministry capable of editing and submitting weekly sermons to air each Saturday morning, but WAVA stepped in and offered support and expertise to ease our anxiety and make airing the broadcast seamless. Ultimately, the scope of our ministry reach has increased and we have received financial support locally and from listeners from as far as West Virginia and New York. Most importantly, our mandate to spread the Gospel is being accomplished and we continue to persevere!"

— **Pastor Tim Dates,** Living Faith Ministries Church, Clinton, MD and host of *It's Your Faith Walk It Out*

"After my arrival as a brand-new senior pastor in 1997 I quickly determined to have a teaching presence on the radio. In 2002 we began broadcasting on the Salem Radio station KFIA, increasing to broadcasting daily in 2006. For over 16 years I have had testimony after testimony of new people coming through our doors telling me that they came because they heard me on the radio. It continues to be both a place for new Christian folks coming into the region to find a new church home as well as an outreach to the unsaved community who can listen in secretly to the Gospel in the privacy of their own car. An unintended blessing of being on KFIA was allowing our church to be a regional encouragement to all the other church leaders who are tirelessly leading their own ministries and blessing their lives."

—**Lance C. Hahn**, senior pastor, Bridgeway Christian Church, Roseville, CA

"I just want to share with you that about 6 years ago I partnered with Salem Media Boston to bring our church ministry beyond our church walls and it surely did. In fact, we needed to add an addition to our church and turn our fellowship hall into an overflow area. If you would like to see similar results, look to Salem Media."

—**Dr. James E. Collins**, senior pastor, Eagle Heights Cathedral in Revere, MA

Cataloging-in-Publication data on file with the Library of Congress

ISBN: 978-1-62157-866-6
ebook ISBN: 978-1-62157-863-5

Published in the United States by
Salem Books, an imprint of Regnery Publishing
A Division of Salem Media Group
300 New Jersey Ave NW
Washington, DC 20001
www.Regnery.com

Manufactured in the United States of America

10 9 8 7 6 5 4 3 2 1

Amplify Your Message

Spreading the Gospel and Growing Your Church Through Radio

Marcus Brown

SALEM
BOOKS
an Imprint of Regnery Publishing

A Note to Pastors

Take a moment to think about noise.

Not the murmuring of a crowded church, or the honking of a grid-locked highway, or the groans of a congested checkout line. The noise you're bombarded with every single day through messages. They originate from so many different sources—smartphones, television, social media—that it's hard to imagine what life would be like if someone wasn't clamoring for your attention everywhere you turn.

So, in this era of so much noise, how do you, as a minister of the Gospel, break through that uproar and deliver a message of hope to the people in your community who need to hear it the most?

There certainly isn't just one answer to that problem, but you've undoubtedly spent a lot of time trying to solve it. In fact, you've probably already experienced how difficult it is to silence the noise that struggles to drown out the peace and love you speak into our world.

We wrote this e-book because we want to share an incredibly effective medium for cutting through that noise. We're talking about Christian radio. It's something we know a lot about here at the Salem Media Group. For decades, we've partnered with local churches and ministries just like yours to break through the four walls of a church building and disperse the message of Jesus everywhere it can go.

Our mission is to offer insight into how Christian radio can help you and your church spread and amplify your message. We recognize that every church is different, and we would never suggest a one-size-fits-all approach to evangelism. But our hope and prayer is that, as you seek the guidance of the Holy Spirit, this e-book will help you develop the most effective communication strategy for your ministry.

—Your Friends at the Salem Media Group

Foreword

In 1875 the Milwaukee Sentinel published a want ad for a local church in search of a pastor. The list of requirements was lengthy and precise. Among others, the ad listed, "Must be willing to preach first-class sermons, and do first-class work, at second-class compensation," and "... should be able to convince all that they are miserable sinners without being offensive." Whereas those job necessities might be knee-slappers to many, pastors aren't laughing. They know the truth; many churches still expect the same things today.

Pastoring isn't for wimps; it's tough work. But, to those whom God has called, we gratefully answered the bell and have given it our all.

Two things tend to keep us awake at night—how to teach and lead with impact, and how to grow the flock.

For decades, America's church attendance has been decreasing—certainly not because of declining population, America has never been more populated. So, where's the breakdown? Oddly enough, media is a big part of the problem.

Perhaps I should explain; several years ago, the Barna Group said, "For the first time in history, more people are getting their spiritual training from the media rather than the church." And, of the different media options, Barna says radio is the runaway winner. As a churchman/pastor, that saddens me—we've been replaced by a technological revolution. However, as a radio executive, I realize God has given us a most advantageous opportunity. In other words, the multitudes are still coming to the church, but they're coming to church via a different vehicle.

Make no mistake about this however; radio must NOT replace the church. That's a core belief within the Salem Media Group. In fact, we believe when God wins a city to Christ, He ALWAYS does it through the local church and never a radio station. And that's where we shine best—we serve as a proven platform to some of the finest churches and ministries in America, it's their message heard loud and clear on our platforms. And we have the uncanny ability to crank up the volume so millions more can hear the claims of Christ.

The document you're looking at—*Amplify Your Message: Spreading the Gospel and Growing Your Church*—was designed to show how quickly and easily a local church can expand its borders and reach. It gives the secrets of feeding a massive audience with a healthy spiritual diet. It unmasks how to make an indelible impact on our culture. And it provides a viable way to grow your church attendance. Proceed with caution but proceed! God called us to spread His word and make disciples. After reading these pages, you'll know how to do that better than ever.

Ron Walters
Pastor
Senior Vice President
Salem Media Group

America's Spiritual Landscape

Before we explore how radio can be an effective tool for your church, we need to examine the spiritual landscape where you minister and the challenges of sharing the Gospel in twenty-first-century America.

As you already know, the religious practices and beliefs of Americans are constantly shifting. In just the last decade, the Pew Research Center reported that the number of Americans who seldom or never attend church worship services has risen to 29 percent, while the number who attend church on a weekly basis has fallen to 37 percent. This means that 63 percent of Americans go to church less than once a month, and almost half of that group never set foot inside a church.

What reasons do these absentees give for missing church? Nearly 24 percent say personal priorities come first, which includes 16 percent who claim they are just too busy.

At the same time, younger Americans—mostly millennials—are rejecting religion entirely. Thirty-five percent of adult millennials (those born between 1981 and 1996) do not identify with any religious label. The percentage of these "nones" is larger than those who identify as evangelicals (21 percent), mainline Protestants (11 percent), or Roman Catholics (16 percent).

The optimists among us might assume that young people rediscover their religious identities as they grow older. But unlike previous generations, 67 percent of millennials have been raised without any religious formation, making it impossible for them to re-examine the nonexistent faith of their childhood.

These statistics may seem disheartening, but we believe the fields of America's spiritual landscape are white for harvest. Even now, we sense that those who currently profess no faith at all are becoming more open to hearing the Gospel. Will you and your church be ready to meet them where they are and present them with the message they need to hear? If so, it's unlikely that such an opportunity will happen inside the four walls of your church.

Does Radio Really Work?

We live in a time when technology moves at a blinding pace. Devices that were cutting edge a few years ago are now obsolete. Just two decades ago, newspapers and the televised evening newscast carried a lot of weight. Today, not as much.

Against a backdrop of constant innovation, is there still a place for a mature medium like radio? You might be surprised to learn that radio has stood the test of time and continues to be an incredibly effective way to reach people.

The Pew Research Center reports, "terrestrial radio continues to reach the overwhelming majority of the public." That fact holds true even with the proliferation of so many new media technologies. Nielsen Media Research—the world's leading organization in media-consumption measurement—has found "traditional AM/FM terrestrial radio still retains its undiminished appeal for listeners—91% of Americans, aged 12 and older, listen to this form of radio."

But you may still think radio isn't really reaching everyone. After all, young people can't be listening to radio, can they? Well, according to Nielsen, "radio's weekly reach among the Millennials across the country is 92 percent." What other medium reaches 92 percent of millennials on a weekly basis?

What's even more exciting is how well radio performs on digital platforms. Take online streaming, for example. According to Edison Research, the percentage of Americans who are twelve years of age or older and have listened to online radio in the past month continues to grow, rising from 53 percent in 2015 to 57 percent in 2016. We'll talk more about the remarkable combination of radio and podcasts a little bit later.

But Does *Christian* Radio Really Work?

When it comes to gathering an audience, Christian radio is broadcasting's dark horse.

According to Lifeway Research, 27 percent of all Americans—Christians and non-Christians— frequently or occasionally listen to Christian radio. All but 1 percent of that group is not affiliated with a church.

Nielsen has refined this research and found that 46 percent of Americans listen to Christian radio every month. One in six listen daily, and 28 percent of non-Christian adults listen monthly. The fastest-growing age group of Christian radio listeners is between the ages of eighteen and thirty-four, recently increasing its consumption by a remarkable 48 percent.

So, why do people listen to Christian radio? We at the Salem Media Group wanted to answer this question, and we commissioned a research project launched by the Harker Research Group to help us. Its findings were astonishing.

Harker discovered that our radio listeners "love the Christian teaching and talk format" and are "very generous and exceedingly loyal." We also learned that their number one reason for giving a financial gift to a church or ministry is a genuine desire "to support your cause."

What makes radio, especially Christian radio, so compelling? We believe this platform's power lies in several of its characteristics.

First, studies have found that most Americans learn best through the spoken word, which makes sense. Almost all of us have learned our most important life lessons through the words of others—parents, teachers, co-workers. And there's a reason why business deals are rarely sealed via e-mail and marriage proposals are hardly ever offered through letters: speaking directly to someone is more compelling than any other form of communication.

Throughout the Bible, we find Jesus and His apostles preaching the Gospel to crowds, large and small, who are eager to listen. But where did their excitement come from? It's likely that a friend or neighbor helped pave Jesus' way into their hearts through a personal testimony. The same is true today. While many of us could have easily picked up a Bible and read it before we became Christians, we probably came to our faith in Jesus after we *heard* someone explain what it means to follow Him.

Second, radio can reach audiences that other mediums can't. Regardless of your outreach efforts, there are many people who will never set foot inside your church. And there are plenty of people who will never read a Christian book. After all, who has time to sit down and pore over a couple hundred pages?

But there are more people who are willing to scan through the radio dial and commit twenty or thirty minutes of their solitary commute to listening to an explanation of the Gospel. That's what makes radio enormously accessible: it's easy, convenient, and free.

Third, radio is surprisingly intimate. Think about it. You typically consume audio content when you're alone in your car, walking your dog, or exercising. Walk down any busy street, and you'll see the same scene repeated block after block: solitary pedestrians with earbuds snugly planted in their ears, listening to something no one else can hear, alone in a sea of people.

That level of intimacy is extremely powerful. Your audience has literally filtered out the rest of the world to listen to what you have to say. Through that car speaker or those earbuds, you are speaking directly to one person and addressing his or her personal fears, needs, and desires. In that moment, a message of Jesus' love for that person can resonate deeply and permanently.

When Radio Meets Digital

One of the most exciting features of Christian radio is how well it transfers to digital platforms. While other traditional forms of communication, like newspapers, have struggled to adapt to the changing media landscape, radio has thrived in this new environment.

Here at Salem Media Group, we have tirelessly sought to provide digital venues for content on our terrestrial radio platforms. We've created limitless access to this content by streaming it through our radio station websites, dedicated mobile apps, and aggregators like TuneIn Radio, iHeart Radio, and our own ChristianRadio.com. The result has been tens of thousands of new listeners who are now accessing this content across these digital platforms.

Our internal data has signaled remarkable growth for such venues over the last few years. The number of sessions started on our streaming platforms grew 19 percent over a period of fourteen months between 2016 and 2017. In that same period, the number of total listening hours (TLH) consumed by our audience increased by 26 percent.

Broader research shows that 80 percent of Americans use an audio streaming service, and that number is over 90 percent among millennials and younger Americans. This is an ideal platform for content that has traditionally lived on terrestrial AM and FM radio stations because such content can be easily streamed on smartphones, tablets, laptops, desktop computers, and smart speakers like Amazon Echo and Google Home. And, luckily, more cars are rolling off the assembly line with audio streaming technology built into their dashboards, which will make it easier for listeners to stream radio content wherever they are.

While audio streaming continues to explode, the podcast revolution is keeping pace. Twenty-four percent of Americans—sixty-seven million people—now listen to podcasts at least once per month. And 71 percent of those who listen on a weekly basis are listening for more than one hour. Podcasts now represent 32 percent of all audio sources for American listeners. In recent years, this number has skyrocketed as media consumption preferences have changed.

More and more consumers want their content on demand. Edison Research has found that 76 percent of Americans agree "the ability to listen to programs whenever [they] want" is very important to them. For audio, this means a growing appetite for podcasts.

Thanks to the universal accessibility of radio on digital platforms, there are no more boundaries keeping you from spreading the Gospel's message to everyone who needs to hear it.

Even the Fastest-Growing Churches Need Radio

The fastest-growing churches in America know that the secret to consistent—and sometimes explosive—growth is to increase their influence in their communities. Through this authoritative presence, spread across a city or region, the Gospel reaches those who most need to hear it. But often, the largest obstacle to extending that outreach is limiting a ministry to the four walls of a church.

Think about what it would take to expand your existing church or ministry to a new site or campus. You would need to acquire a building or raise a significant amount of money to construct one. Once the building was secured, it would need to be outfitted to suit your ministry needs with furniture, technology, audio visual capabilities, and new or reallocated staff. All of this would require an enormous amount of time and resources. Even if you began such an initiative today, it would likely take months, if not years, to come to fruition.

Now, consider the possibility of radio becoming your next campus. All of a sudden, those obstacles of time and resources disappear. The timeline of months or years shrinks down to weeks or days. Rather than investing in a physical location that reaches several hundred or a thousand people, you can more efficiently invest in something that can reach tens of thousands, or even hundreds of thousands, of people throughout your city and across the world.

If you already have multiple campuses spread throughout your community, you know one of the risks of a multi-site church is the disconnect that can develop between your congregations. Thankfully, you can bridge that gap with radio. Your program, broadcast across your community and distributed online through audio streaming and podcasts, will unite every person who calls your church home, regardless of which campus they attend. Through these broadcasts, you can easily build a community across multiple campuses and keep your members in touch with the heartbeat of your ministry between weekend services.

Whether your church has one site or many, launching a radio campus can strengthen the bonds that tie together a thriving ministry and open the floodgates for new members to discover and join your church.

Can't I Just Go Online and Do It Myself?

After all, anyone can start a podcast, right? Well, you and your church certainly could launch an online broadcast without help, but reaching your audience would be a steep, uphill climb. Right now, there are over 250,000 separate podcasts on iTunes and almost 100,000 streaming options on TuneIn Radio, one of America's most popular streaming apps.

So, striking out on your own would be like opening a kiosk in the middle of a mall with several hundred thousand other stores. You'd be lucky to have more than one or two customers stop by every day, assuming they could find you.

Thankfully, radio stations need churches to reach the mission field. Radio has a platform, but the church has the message. This is why the relationship between a local Christian radio station and its community is so critical. It truly is a partnership.

Christian radio stations see the communities they serve as mission fields, and they want to "adopt" these communities so that the Gospel can reach them. But they simply cannot do it alone. They understand that their role is to support the work that's already happening inside local churches by amplifying it across their communities.

The reason this partnership works so well is because Christian radio stations and churches understand their roles in the bigger mission. You and your church can focus on your core competencies and calling while a Christian radio station's pre-established platform provides the technical infrastructure and reach to get your message—His message—heard.

How Does It All Happen?

Developing a radio strategy for your church can seem daunting, but one of the benefits of a partnership between your church and a local Christian radio station is that the burden isn't on you to make it all happen.

Once you decide what kind of program you will broadcast—many pastors simply repurpose sermon series or messages from church conferences with a common topic—your Christian radio station will help you develop production elements that will amplify your message and reach your audience. Typically, these elements include a thirty- or forty-five-second opening, middle, and closing. Your entire broadcast, including its production elements, will usually be twenty-five minutes long.

The opening contains a brief welcome to your listeners, a brief introduction of the episode's message, and the names of your program, its host, and your church. The middle, which is the briefest production element, interrupts the episode to remind listeners which program they are listening to. It may also contain a short call to action, which we'll discuss later. The closing provides a more definitive call to action, along with information about how to contact your church, attend its weekend services, and learn more about its upcoming events.

Production elements can be recorded by a pastor or another staff member who is the designated "voice" of the program, or they can be recorded by a professional voice-over artist, provided by the radio station. These elements should be updated regularly to reflect the specific content of each episode and the most up- to- date information about your church's activities.

When you've created a template for your program's production elements and decided who will record them, you will be able to convert your existing sermons or messages into a broadcast. But there are other kinds of programs that may suit your church better. These require more preparation and ongoing production than repurposing existing content from your ministry.

One alternative program is a more traditional talk show, featuring one or two hosts who engage in a focused conversation throughout each episode. To start a talk show, you need to decide who will host the program and how you will select topics for each episode. As a rule of thumb, block out *at least* one hour of off-air preparation for every thirty minutes of on-air time. When the show is ready to be produced, one of the two hosts will deliver the program's

opening and then ask the other host questions. Without even trying, this style feels accessible and relatable to most listeners, especially those who have never heard a sermon before.

If your church does not already have the necessary equipment to record a radio program, you may need to use a production studio at your local Christian radio station to record each episode of your program. You are always able to consult with your station to determine the best production option for your program.

Another type of program you may consider producing is an interview-style program. This type of broadcast features a host who interviews a guest on a new topic each episode. Interviewees can be staff members of your church or members of your congregation. They could also be Christian authors, speakers, or experts on the topic discussed during their interview.

This type of program is best suited for ministries that use the wisdom and experiences of others to communicate their message. For example, on some interview-style shows, pastors interview congregation members about their personal testimonies. Other programs allow pastors to interview their staff members about topics like marriage, raising children, or dealing with grief or loss.

The last program we will examine is the "call-in" show, featuring a host who addresses call-ins from listeners with questions about an episode's topic. This type of show is always hosted live, which makes it highly interactive. In fact, one of its greatest strengths is making direct contact with listeners seeking guidance. You're able to immediately answer their questions and follow up with them off-air when possible.

Call-in shows also allow listeners to anonymously ask questions about aspects of their lives they would otherwise not feel comfortable or safe discussing. It's much easier to ask a question over the phone than it is to walk into a pastor's office and ask that same question face to face. In some cases, you will have an incredible opportunity to minister to people in unstable or desperate situations.

A challenge to broadcasting a program like this is finding a host who can handle the types of questions that may be asked. We've heard pastors answer call-in queries about everything from misconceptions about the Bible and ethics to doubts about marriage and parenting. It isn't easy to think on your feet and quickly offer thoughtful responses to difficult questions like that. Obviously, a call- in show also requires callers, and it can take several weeks for listeners to become comfortable enough with your program to call you. In the meantime, you may need to solicit questions from other sources, like social media.

As you think about what kind of radio program best suits your ministry, also consider the audience you are trying to reach. You may target non-believers who have little experience with church and are not completely open to hearing a sermon. Perhaps you want to reach believers who don't have a regular church home or are trying to grow spiritually through a deeper understanding of the Bible. Sifting through these considerations will help you determine which broadcasting format to pursue, and your local Christian radio station can assist you throughout this process.

Getting On the Air

Once you've selected the best format for your program, you need the necessary buy-in from your ministry partners to get your program on the air. Often, this means obtaining your church board's approval, and that can be one of the greatest challenges to launching a radio ministry.

Whoever needs to approve your radio ministry, always remember that they are called to be the best possible stewards of your church's resources. If you've ever sat on a church board, you know how many difficult decisions have to be made about allocating limited resources. Every dollar spent in one area cannot be spent in another. Like every institution, churches must prioritize their resources.

While every church board is different, there are two questions you should always answer before you propose a new church initiative like a radio ministry: (1) what's the return on this investment? and (2) will this ministry be sustainable? These are both very important questions that we will attempt to address here.

At first, it may seem odd that a church board would want to know the predicted return on an investment. It's easier to imagine the CFO of a large company asking such a question. But it's probably not uncommon to hear someone at a church board meeting ask, "What are we getting for our money?" It's an appropriate question whether we're talking about spending money to pave a church parking lot, hire a new children's minister, donate to missionaries, or launch a radio ministry.

Depending on your ministry strategy, that first question can be answered in many different ways. For some churches, the return is clear—people coming to Christ and joining your church family. Many radio ministries focus on attracting new individuals and families to the church, and this strategy can yield a high return if it's executed with that mission in mind. More and more frequently, visitors are introduced to a church before they walk through its door. This introduction is often a message they heard on the radio. If you're ready to welcome new members into your congregation, a radio ministry can be a very effective way to fuel that growth.

You should also think about the spiritual return on the investment you're asking your church to make. This gradual return may be tough to measure, but it's no less important than a more immediate one.

As you share your ministry through radio, you will reach people from all walks of life. Your message may be exactly what they need to hear to walk away from destructive habits, to recommit to their marriages, to rededicate their lives to Christ, or to simply find the strength to get out of bed that day. In each of those situations, you may not meet the people you've helped. They may never step foot in your church or drop a check in your offering plate. But helping Christ's kingdom take root in every corner of our world is one of the greatest spiritual returns a ministry can receive.

When you plant seeds, it's impossible to know which will flourish and which will not. But you and your church may be called to plant those seeds in faith, knowing there will be a harvest someday.

So, what about that second question—can a radio ministry be sustainable? In other words, will it always rely on the church's resources or can it eventually stand on its own feet? This question touches on something we mentioned earlier: a call to action. Every episode of your broadcast gives you an opportunity to ask your listeners to do something worthwhile.

The simplest action you can ask your listeners to do is pray—for your ministry, for the community, and for what the Lord may be calling them to do in response to the message they hear on your program.

Occasionally, you may ask more from your listeners in exchange for a resource, like a book, DVD, or digital download. Your request can be as small as them sending you their mailing or e-mail address so that you can add their contact information to a larger database for future communication. You may also offer your listeners an opportunity to give a financial gift (of a suggested or fixed amount) to help support your ministry.

Of course, to participate in this interchange, you need to develop resources for your listeners. Perhaps you already have printed material—books, booklets, tracts, or pamphlets—that can be shared easily. Or maybe you have recorded media, like DVDs, CDs, or downloadable MP3s of sermons, teaching series, or special conferences. You can also convert sermons, sermon series, blog posts, or substantial social media posts into a booklet or downloadable e-book.

Making this exchange between you and your listeners as meaningful as possible will help you develop a network of followers who are led by the Lord to provide ongoing support for your ministry. Sometimes, listeners immediately feel called to offer financial gifts. But it is more common for an emerging network to feel called to give after weeks, months, or years of regular and consistent contact with the blessings of your ministry.

The success of your program depends on the growth of this network, so it is crucial for you and your church board to think through these questions of return and sustainability. You must have a clear, shared vision for your program, and everyone must understand what is required to make it a reality. This process can take time and should always be undergirded with prayer.

More than one pastor has shared a story about a skeptical church board that was reluctant to support a radio ministry until a miraculous outpouring of generosity turned the impossible on its head. That kind of response to ministry only happens when the process is sustained by prayer.

Promoting Your Program

Before you launch your program, make sure people will know about it. There are countless promotional strategies you could use to spread the word about your program, but we think there are only a few that no Christian-radio marketing plan should be without.

Your church is the most obvious place where you should start promoting your program. Let your congregation know your program is on the air and available to download. Only you know the best way to reach your church, whether it's through a bulletin insert, signage around your building, a slide in your weekly announcement loop, or takeaway items members can pick up in your building. And don't hesitate to ask your church to share the program with a friend. One of the easiest ways for your congregation to invite a friend or neighbor to church is to first invite them to listen to your broadcast.

Be precise when you circulate details about your program. Where can your congregation download your podcast and how can they subscribe to it? What's the broadcast's radio station frequency? On which days and at what times should they listen? Why should they listen?

Regarding external marketing, promoting your program on your social media platforms is the easiest first step you can take. Keep in mind that these platforms reward content over blatant self-promotion, so make sure you have enough time to create advertisements that stand out.

Rather than simply telling people to listen on Facebook, it's always better to pique their interest with a pull-quote from your program that you can incorporate into a unique post. Don't forget to include information about how your followers can hear the rest of the message either by listening live or by downloading the episode the quote came from.

You can also grab attention on social media by posting provocative questions or shocking statistics that relate to an upcoming episode of your program. Another powerful way to generate interest in your program is to tell people what they've been missing. Try posting listener feedback about how your broadcast has impacted the lives of your audience.

E-mail subscribers are often the most loyal followers, so make sure they are also looped into updates about your program. You can even e-mail them an invitation to create a calendar event for your program every day or every week.

Outside of social media, there are many other tactics for spreading the word about your program. Some churches pay for online ads that look like website banners or are triggered by certain browser searches. One savvy church used search terms related to addiction, depression, and pornography usage to display ads about their radio program. It's a great way to offer hope to someone who might be crying out online.

Billboards can also be a great way to generate listeners. Often, drivers will turn on their radios if they see a compelling message from a captivating ministry on the road.

We've also found that printed material can direct a lot of traffic to your program. Some churches have redesigned their staff's business cards to include details about their radio programs right next to their weekly service times. Others distribute print collateral that can be displayed in local coffee shops, schools, businesses, or other community hotspots. This kind of collateral can be formatted into larger items, like posters or flyers, or smaller ones, like "touch cards"—business-card-sized items that only promote your radio program.

There is no silver bullet when it comes to marketing. But it's important to remember that once your program is on the air, it's like a snowball at the top of a hill. It needs to build momentum to grow. Promotion is that momentum, so don't hesitate to give it a few helpful pushes along the way.

So, What's Next?

You know how incredibly difficult it is to spread the Gospel, especially in America's current spiritual climate. We've illustrated the challenges you face and demonstrated how radio— coupled with powerful digital platforms—can help you overcome them and fulfill your calling. We've also shown that reaching your audience requires a true partnership between a local church and a local Christian radio station that work together for a common mission.

The prospects may seem exciting, but the task can feel daunting. This is why each Salem Christian radio station has a local ministry director who is prepared to take the first step with you, to determine what the Lord may have in store for your church. Ask your local ministry director about the twenty-one steps you need to take to make radio ministry a reality for your church.

Before we conclude, we want to share the final step with you because, in many ways, it's the most important. It's simple: make sure you and your church pray that the Lord is in this decision and remains in the process at all times. If you keep Him and His Word at the center of your ministry, your plans of expanding it through radio will grow exponentially. Your friends here at Salem want God's best for you, your ministry, and ultimately your efforts for Christ's Kingdom. We would love the opportunity to come alongside you in any way we can for the sake of the Gospel.

Printed in the United States
By Bookmasters